Discovery EDUCATION | SCIENCE TECHBOOK

California Unit 4

Motion and Change

To obtain permission(s) or for inquiries, submit a request to:
Discovery Education, Inc.
4350 Congress Street, Suite 700
Charlotte, NC 28209
800-323-9084
Education_Info@DiscoveryEd.com

ISBN 13: 978-1-68220-532-7

Printed in the United States of America.

3 4 5 6 7 8 9 10 CWM 27 26 25 24 23 22 B

Acknowledgments

Acknowledgment is given to photographers, artists, and agents for permission to feature their copyrighted material.

Cover and inside cover art: Lee Yiu Tung / Shutterstock.com

Table of Contents

Unit 4

Concept 4.1

Concept 4.2

Unit Project

Grade K Resources

Dear Parent/Guardian,

This year, your student will be using Science Techbook™, a comprehensive science program developed by the educators and designers at Discovery Education and written to the California Next Generation Science Standards (NGSS). The California NGSS expect students to act and think like scientists and engineers, to ask questions about the world around them, and to solve real-world problems through the application of critical thinking across the domains of science (Life Science, Earth and Space Science, Physical Science).

Science Techbook is an innovative program that helps your student master key scientific concepts. Students engage with interactive science materials to analyze and interpret data, think critically, solve problems, and make connections across science disciplines. Science Techbook includes dynamic content, videos, digital tools, Hands-On Activities and labs, and gamelike activities that inspire and motivate scientific learning and curiosity.

You and your child can access the resource by signing in to www.discoveryeducation.com. You can view your child's progress in the course by selecting Assignments.

Science Techbook is divided into units, and each unit is divided into concepts. Each concept has three sections: Wonder, Learn, and Share.

Units and Concepts Students begin to consider the connections across fields of science to understand, analyze, and describe real-world phenomena.

Wonder Students activate their prior knowledge of a concept's essential ideas and begin making connections to a real-world phenomenon and the **Can You Explain?** question.

Learn Students dive deeper into how real-world science phenomenon works through critical reading of the Core Interactive Text. Students also build their learning through Hands-On Activities and interactives focused on the learning goals.

Share Students share their learning with their teacher and classmates using evidence they have gathered and analyzed during Learn. Students connect their learning with STEM careers and problem-solving skills.

Within this Student Edition, you'll find QR codes and quick codes that take you and your student to a corresponding section of Science Techbook online. To use the QR codes, you'll need to download a free QR reader. Readers are available for phones, tablets, laptops, desktops, and other devices. Most use the device's camera, but there are some that scan documents that are on your screen.

For resources in California Science Techbook, you'll need to sign in with your student's username and password the first time you access a QR code. After that, you won't need to sign in again, unless you log out or remain inactive for too long.

We encourage you to support your student in using the print and online interactive materials in Science Techbook on any device. Together, may you and your student enjoy a fantastic year of science!

Sincerely,

The Discovery Education Science Team

Unit 4

Motion and Change

Cars, Trains, and Boats

Quick Code: ca750s

Cars and trucks seem to move without another object touching them. Other objects stay still until another object touches them. By the end of this unit, you will be able to explain why objects like cars, trains, sleds, and boats move. You will know how to change the way things move.

Video

Cars, Trains, and Boats

Think About It

Look at the photograph. **Think** about the following questions:

- What causes motion?
- How can we control motion to meet our needs?

Fair Ride

Design Solutions Like a Scientist

Quick Code: ca834s

Hands-On Engineering: Investigating Motion

In this activity, you will design a solution to move a heavy box from a low place to a high place.

Moving Boxes

SEP Constructing Explanations and Designing Solutions

SEP Planning and Carrying Out Investigations

SEP Asking Questions and Defining Problems

CCC Cause and Effect

Ask Questions About the Problem

You are going to design a solution that will help you to move a heavy box. You will move the box from a low place to a high place. How will you do this?

HANDS-ON ENGINEERING

Write some questions you can ask to learn more about the problem. As you work on activities throughout the unit, **write** down answers to your questions.

CONCEPT
4.1

Push and Pull

Student Objectives

By the end of this lesson:

- [ ] I can look and study to tell whether an object moves because of a push or a pull.
- [ ] I can test how a strong push or pull moves an object differently than a weak push or pull.
- [ ] I can test what changes the speed of an object.

Key Vocabulary

- [ ] energy
- [ ] force
- [ ] gravity
- [ ] living
- [ ] motion
- [ ] nonliving

Quick Code: ca753s

Activity 1

Can You Explain?

What causes a skateboard to move?

Quick Code:
ca755s

Activity 2

Ask Questions Like a Scientist

Quick Code: ca756s

Skateboarding

Watch the video. **Think** about the different ways the skateboard moves.

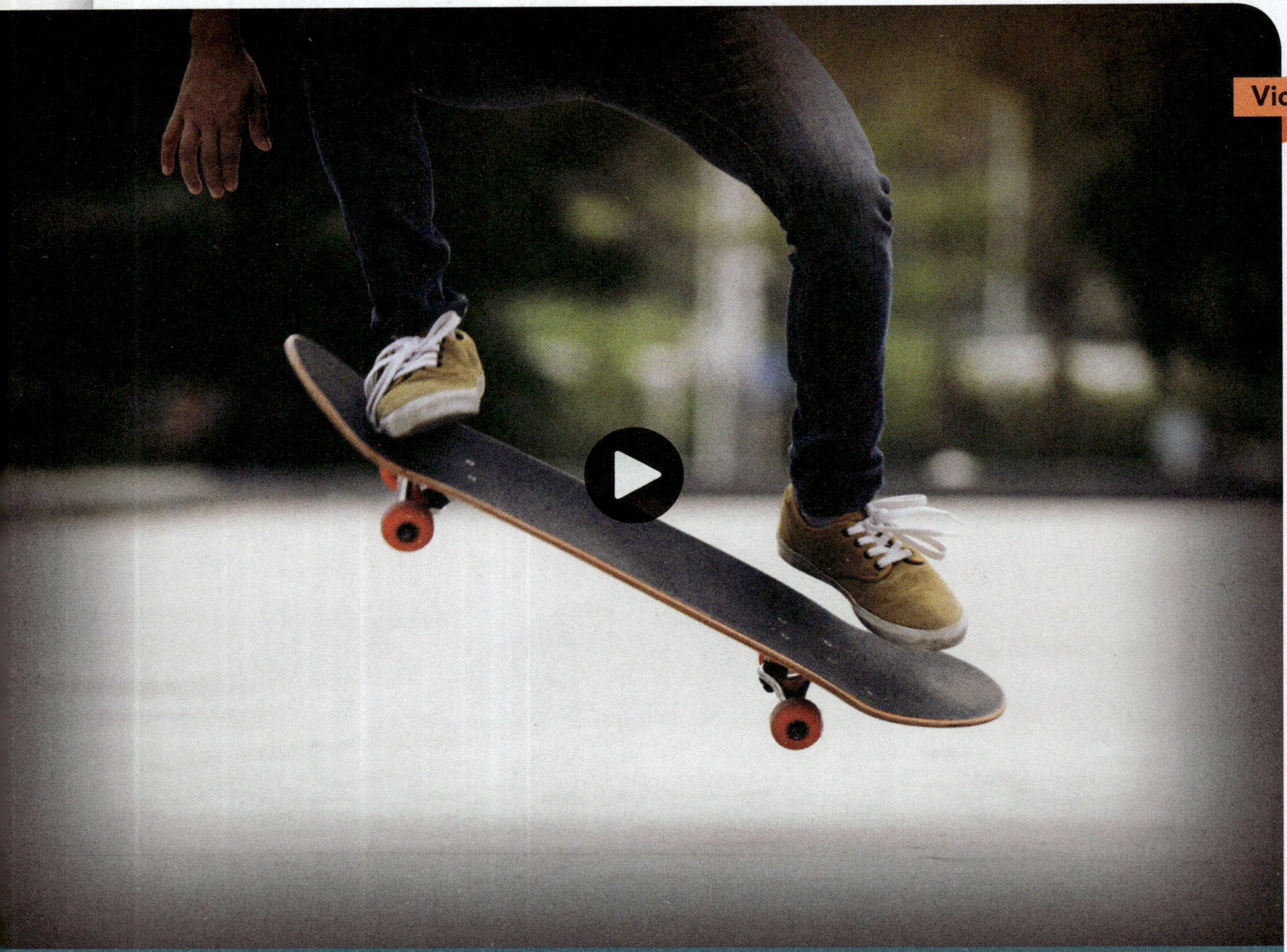

Let's Investigate Skateboarding

Talk Together

Now, talk about the questions you might ask the skateboarder about how he makes the skateboard move.

SEP Asking Questions and Defining Problems

Activity 3

Analyze Like a Scientist

Quick Code: ca758s

How Do Things Move?

Read the text. **Think** about ways that things move.

Read Together

How Do Things Move?

You are a **living** thing. You move all day long.

You also help move objects in your classroom or on the playground.

What objects did you move today?

Read Together

What Things Move?

Living beings, such as people and animals, can move.

Movement

Nonliving objects, such as balls and blocks, can also move.

What things do you see around you? How do these things move?

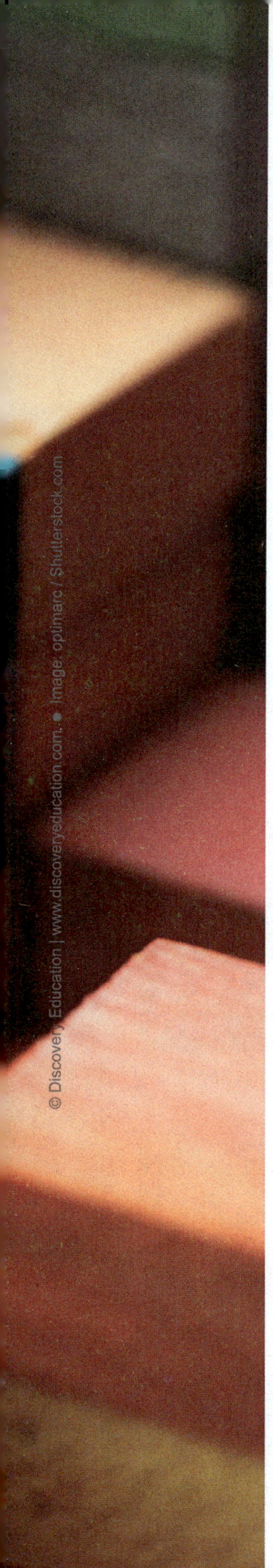

Activity 4

Analyze Like a Scientist

Quick Code: ca759s

What Things Move?

What things move by themselves? What things need help moving? **Write** your ideas on the chart.

Things That Move

Moves by Itself	Needs Help Moving

Activity 5

Evaluate Like a Scientist

Quick Code: ca760s

What Do You Already Know About Push and Pull?

How Do Objects Move?

Look at the pictures of moving things.

Wagons

Stroller

Bicycle

Sled

Lawnmower

Playing with Dogs

Which pictures show something that moves by pushing? Which show something that moves by pulling?

Write your answers in the chart.

By Pushing	By Pulling

Thinking of Moving Things

Motion describes how everything moves.

Both living and nonliving things can move.

Living beings, such as people and animals, can move. Nonliving objects, such as balls and blocks, can also move.

Talk Together

Motion describes how everything moves.
Take turns naming words that describe motion.

Things That Move

Which of the things in the pictures move on their own? Which only move when pushed or pulled? **Draw** a line from each picture to the correct words.

Bird

Basketball

Tennis Ball and Racket

Horse

Moves on Its Own

Moves When Pushed or Pulled

Pumpkin Cart

Dog

Golf Club and Ball

Cat

What Words Can We Use to Talk About How Something Moves?

Activity 6

Observe Like a Scientist

Quick Code: ca764s

Things Moving

Watch the video. **Look** for what makes the rocking chair move.

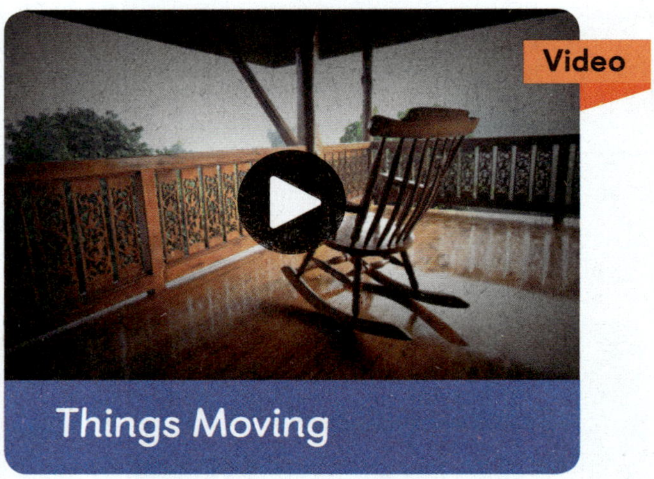

Things Moving

SEP Constructing Explanations and Designing Solutions

CCC Cause and Effect

Talk Together

Talk together about what makes the rocking chair move.

Draw a picture of somebody moving an object.

Activity 7

Analyze Like a Scientist

Quick Code: ca765s

Living and Nonliving Things Move

Read the text. **Circle** the words that describe motion.

Read Together

Living and Nonliving Things Move

Motion describes how things move.

Both living and nonliving things can have motion.

SEP Obtaining, Evaluating, and Communicating Information

CCC Cause and Effect

Living things like people can run, walk, crawl, bend, jump, or hop.

Nonliving things can roll, bounce, or slide.

Things can move fast and slow.

Children Running

Activity 8

Think Like a Scientist

Quick Code: ca766s

Did It Move?

In this activity, you will describe how objects moved.

What materials do you need?
(per group)

- Cardboard
- Crayons
- Pencils
- Markers
- Eraser
- Paper clips

SEP Constructing Explanations and Designing Solutions

CCC Cause and Effect

What Will You Do?

Put your objects on a desk.

Hold a piece of construction paper in front of the objects.

Move some of the objects behind the paper.

Have your partner tell which objects moved.

Which objects moved?

How did the objects move?

Think About the Activity

How did you know that the objects moved? **Write** or **draw** your answer.

What Are the Reasons Something Might Start or Stop Moving?

Activity 9

Observe Like a Scientist

Quick Code: ca767s

Pushes and Pulls

Watch the video. **Look** for changes that are caused by pushes and pulls.

Pushes and Pulls

SEP Obtaining, Evaluating, and Communicating Information

Write down things that need a push to move.

Write things that need a pull to move.

Pushes	Pulls

Activity 10

Analyze Like a Scientist

Quick Code: ca768s

Playground Motion

Think about what you read. Then, add objects to the Pushes and Pulls table.

Read Together

Playground Motion

A **force** makes something move. A force can be a push or a pull.

Seesaw

You can make a ball move by pushing it.

You can bring an object closer to you by pulling it.

CCC Cause and Effect

Activity 11

Observe Like a Scientist

Quick Code: ca769s

Measuring Things in Motion

Watch the video. **Look** for objects that are pushed and objects that are pulled.

Measuring Things in Motion

Talk Together

Now, talk together about how pushes and pulls make things move fast or slow.

SEP Obtaining, Evaluating, and Communicating Information

Activity 12

Investigate Like a Scientist

Quick Code: ca770s

Hands-On Investigation: Cars in Motion

In this activity, you will look at what causes a change in motion.

Make a Prediction

What do you think will happen when you push the toy car gently? What do you think will happen when you push the toy car hard? **Write** or **draw** your prediction.

What materials do you need?

(per group)

- Toy trucks
- Metersticks
- Toy trucks, pull back
- Pull back car

HANDS-ON INVESTIGATION

What Will You Do?

Give a toy car a gentle push. **Record** how far it travels before it stops.

Give a toy car a hard push. **Record** how far it travels before it stops.

Repeat the same pushes with a different car or truck. **Record** how far each travels before it stops.

SEP Planning and Carrying Out Investigations

SEP Using Mathematics and Computational Thinking

CCC Cause and Effect

Car	Type of Push	Distance Traveled
Car 1	Gentle	
Car 1	Hard	
Car 2	Gentle	
Car 2	Hard	

Think About the Activity

Draw how each type of push caused a change in motion.

Gentle Push

Hard Push

Was your prediction right about the gentle push?

Was your prediction right about the hard push?

How does the distance your car rolled with a hard push compare to the distance it rolled with a gentle push?

The car I pushed hard moved a ______ distance. The car I pushed gently moved a ______ distance.

What Causes Something to Move on Its Own?

Activity 13

Observe Like a Scientist

Quick Code: ca771s

Moving a Pet Stone

Watch the video. **Think** about why the stone cannot move by itself.

Moving a Pet Stone

SEP Obtaining, Evaluating, and Communicating Information

CCC Cause and Effect

Talk Together

Now, talk together about **energy**. Talk about why objects need energy to move.

Read Together

Some Things Move by Themselves

Some things can move by themselves. Animals, people, and automobiles have energy inside them. This energy allows them to move on their own.

Some things move because another object with energy makes them move. A ball moves when your hand pushes it.

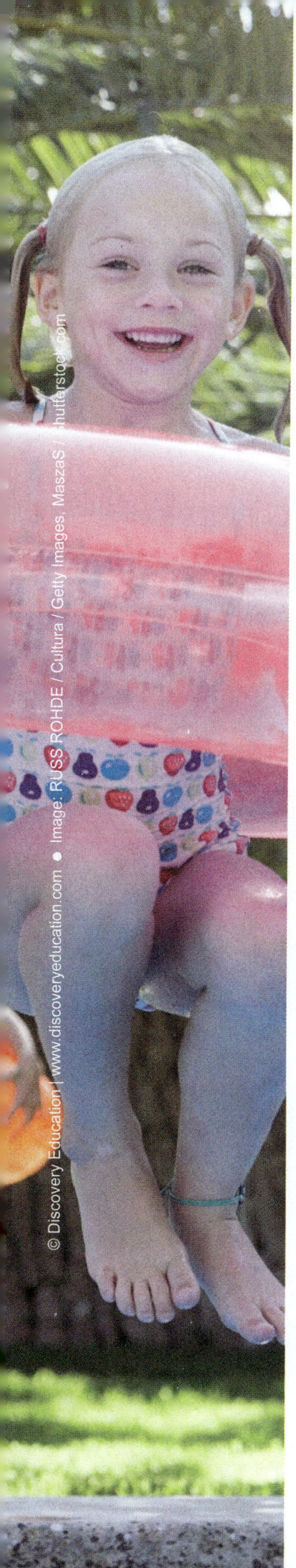

Activity 14

Analyze Like a Scientist

Quick Code: ca772s

Some Things Move by Themselves

Look at the picture.

Circle things in the pool that can move on their own.

Children in Pool

CCC Patterns

CCC Cause and Effect

Activity 15

Evaluate Like a Scientist

Quick Code:
ca774s

Using Energy

Objects use energy at different times. **Look** at the pictures. **Circle** the ones that show objects that are using energy. **Draw** a square around the ones that show objects that are not using energy.

Books

Bicycle

Train

Basketball and Hoop

Hay Bales

Boat

SEP Constructing Explanations and Designing Solutions

CCC Patterns

Why Does Something Fall Down?

Activity 16

Observe Like a Scientist

Quick Code: ca775s

Gravity

Gravity causes things to fall down. Gravity is a force. It pulls objects toward the ground. When you go down a slide, you are being pulled by gravity.

Watch the video. **Think** about how things you can see are being pulled by gravity.

Gravity

SEP Planning and Carrying Out Investigations

CCC Cause and Effect

Draw a picture of something that is being pulled by gravity.

Activity 17

Observe Like a Scientist

Quick Code: ca776s

Making Things Move

Go online to **explore** push and pull forces, gravity, and how fast something moves. Then, **answer** the questions.

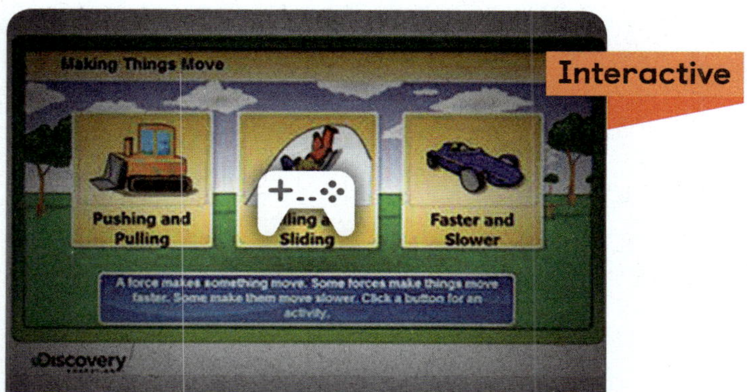

Making Things Move

Was the force on the motorboat a push or a pull?

SEP Planning and Carrying Out Investigations

CCC Cause and Effect

What force causes you to fall down when you slip on a wet floor?

What force helps you move a heavy box down a ramp?

Which car was faster: the blue car or the red car?

How do you know which car was faster?

Activity 18

Investigate Like a Scientist

Quick Code: ca777s

Hands-On Investigation: High and Low, Fast and Slow

In this activity, you will look at how gravity pulls an object down a ramp.

Make a Prediction

What will happen when the ramp is low? **Write** or **draw** your prediction.

What will happen when the ramp is high? **Write** or **draw** your prediction.

What materials do you need?

(per group)

- Stiff and sturdy piece of cardboard at least 2 feet long and 5 inches wide, 1
- Connecting cubes
- Metersticks
- Clear tape
- Masking tape
- Stopwatch
- Toy trucks

What Will You Do?

Build a low ramp, a medium ramp and a high ramp.

Let your truck roll down each ramp.

Measure the time it takes your truck to go down each ramp. **Write** the time.

Height of Ramp	Time It Took Truck to Travel to the End
1 Block	
2 Blocks	
3 Blocks	

SEP Planning and Carrying Out Investigations

SEP Using Mathematics and Computational Thinking

CCC Cause and Effect

Think About the Activity

Draw a picture of the ramp with the fastest truck.

Draw a picture of the ramp with the slowest truck.

When did your truck go fastest down the ramp?

When did your truck go slowest down the ramp?

Activity 19

Evaluate Like a Scientist

Quick Code: ca778s

Will the Ball Fall?

Look at each picture.

Decide if the ball will keep falling or if the ball will not fall.

Draw a line from each picture to the correct words.

SEP Engaging in Argument from Evidence

CCC Cause and Effect

Soccer Balls

Basketball and Hoop

Playing Catch

Baseball and Glove

Ball Will Fall

Ball Will Not Fall

Activity 20

Solve Problems Like a Scientist

Quick Code: ca782s

Motion Maze

Read Together

Have you ever played with a maze game? Have you built a marble run? What causes the round object to go through a maze or down a marble run?

SEP Planning and Carrying Out Investigations

CCC Cause and Effect

Sometimes, mazes use gravity to move a ball through the tracks. The ball is placed at a higher location and runs down through the maze to the end.

Other times, the maze is flat and the ball needs extra energy to go through the maze.

Maze

What forces will cause a round object to travel through a maze?

Submit a Maze

For this project, you will make a maze from cardboard and craft sticks. Then, you will move a small round object through the maze.

Ask Questions About the Problem

What questions do you have about this activity? What do you need to know to make a marble maze? **Write** your questions below.

What materials do you need?
(per group)

- Camera (to take video/picture of mazes)
- Cardboard box with sides (shoebox tops or other)
- Craft sticks
- Disposable straws
- Marbles, 1 in.
- Masking tape

What Will You Do?

1. **Use** your cardboard box as the container in which you make your maze.

2. **Plan** where you want to glue craft sticks to create a path for the object to travel through. **Draw** a picture of where you will place your craft sticks within the box. Your maze should have four or more turns for the object to make.

3. **Apply** glue to the sides of a craft stick and place it in the box. **Glue** one craft stick at a time to create a maze path through your box.
4. **Wait** for your glue to dry thoroughly before completing the next part of the project. While you are waiting, **take a photo** of your maze. You will upload the photo of the maze at the end of the project!
5. When your maze is completely dry, you and your partner will take turns blowing through your own straws to push the object through the maze. Try blowing gently on the straw. What do you notice? Try blowing hard on the straw. What do you notice? Does the object travel through the maze more easily if you blow softly or if you blow hard?
6. **Talk** to your partner about how the object went through the maze. **Use** directional words like near and far, left and right to describe what happened when you blew through your straw to move the object.

Think About the Activity

Draw the round object moving in the maze.

Round Object Moves Right →
Round Object Moves Left ←

What words describe how the round object moved in the maze?

How did gravity move the round object in the maze?

Activity 21

Record Evidence Like a Scientist

Quick Code: ca779s

Skateboarding

Now that you have learned about pushes and pulls, look again at the Skateboarding video. You first saw this in Wonder.

Let's Investigate Skateboarding

Talk Together

How can you describe Skateboarding now?
How is your explanation different from before?

SEP Constructing Explanations and Designing Solutions

Look at the Can You Explain? question. You first read this question at the beginning of the lesson.

Can You Explain?

What causes a skateboard to move?

Now, you will use your new ideas about skateboarding to answer a question.

1. **Choose** a question. You can use the Can You Explain? question, or one of your own. You can also use one of the questions that you wrote at the beginning of the lesson.

My Question

2. Then, **use** the sentence frames on the next page to answer the question.

I know that the skateboard

because

When the skateboard moved forward, it was because a force

The skateboarder used energy to

The force of gravity pulled

on the skateboard.

STEM in Action

Quick Code: ca780s

Activity 22
Analyze Like a Scientist

Helping People Move

Read about helping people move. Then, **try** the activity.

Read Together

Helping People Move

Can you think of jobs that help people move? Coaches and P.E. teachers help kids move.

A Fast Ball

Some coaches help kids throw baseballs farther or run faster. P.E. teachers help kids stay healthy by teaching them different ways to move.

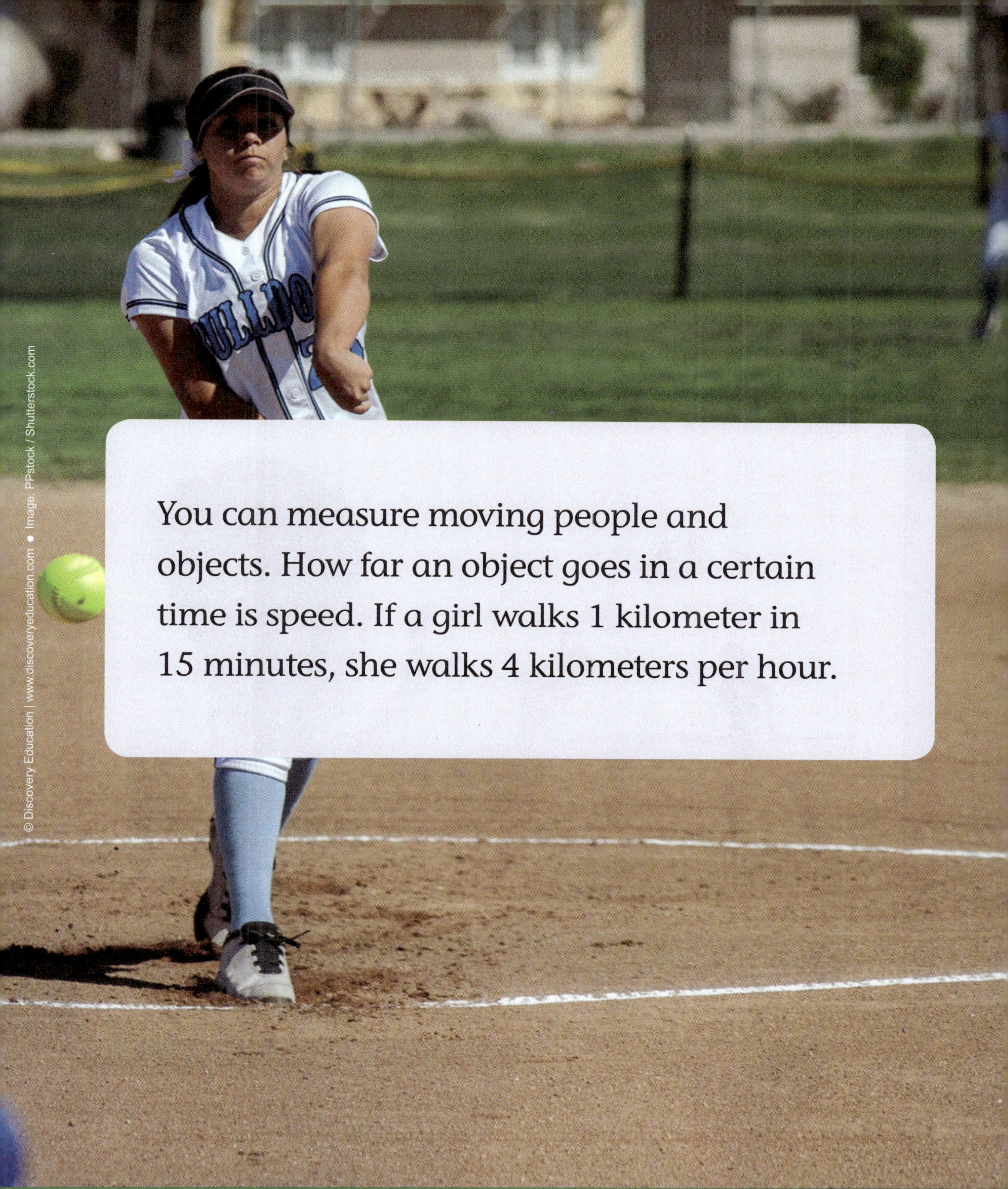

You can measure moving people and objects. How far an object goes in a certain time is speed. If a girl walks 1 kilometer in 15 minutes, she walks 4 kilometers per hour.

Throwing a Ball Harder

Some people make new products, or they make old products better. Imagine you want to make a better football. You have to test it to see how it moves.

Testing a Football

You threw the football to a friend, and it went 20 yards. How could you change your force to throw to friends at different distances?

Match each amount of force to how far the football will go.

	5 yards
Less force	
	10 yards
More force	
	20 yards
Same amount of force	
	30 yards

Activity 23

Evaluate Like a Scientist

Quick Code: ca783s

Review: Push and Pull

Think about what you have read and seen. What did you learn?

Draw what you have learned. Then, **tell** someone else about what you learned.

SEP Constructing Explanations and Designing Solutions

Talk Together

Think about what you saw in Get Started.
Use your new ideas to discuss push and pull.

CONCEPT 4.2

Controlling Motion

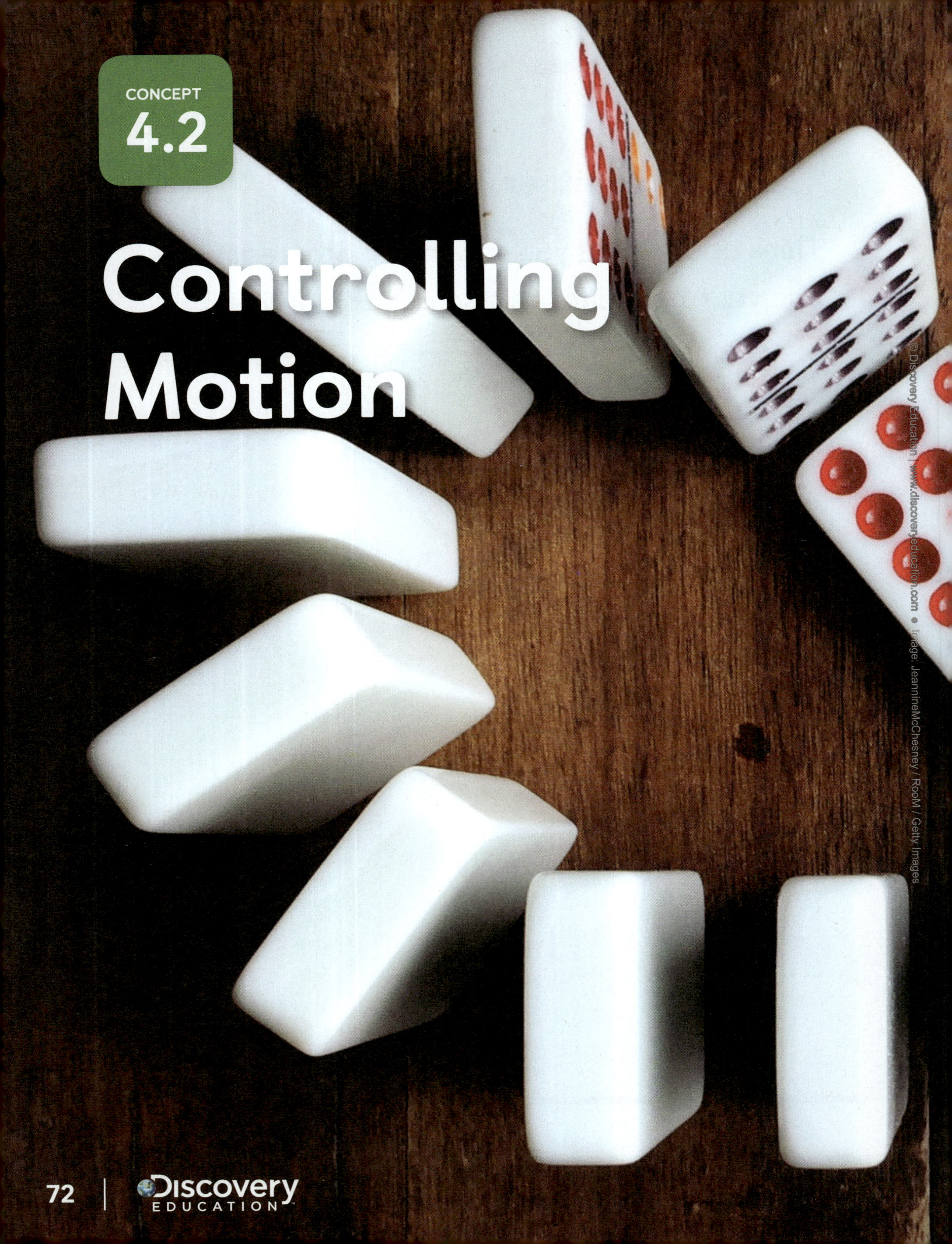

Student Objectives

By the end of this lesson:

- ☐ I can test how a pull or push can change the speed of an object.
- ☐ I can test how a pull or push can change the direction of an object.
- ☐ I can solve problems by changing an object's motion and make my solutions better.

Key Vocabulary

- ☐ balanced
- ☐ direction
- ☐ force
- ☐ friction
- ☐ function

Quick Code: ca790s

Activity 1

Can You Explain?

How can we change the speed and direction of a bowling ball?

Quick Code:
ca791s

Activity 2

Ask Questions Like a Scientist

Quick Code: ca793s

A Moving Bowling Ball

Let's Investigate a Moving Bowling Ball

SEP Asking Questions and Defining Problems

There is a story behind every picture. **Look** at the picture of the bowling ball and bowling pins.

Talk Together

Now, talk about what questions you would ask to find out why some pins moved and some did not.

Activity 3

Analyze Like a Scientist

Quick Code: ca794s

Moving Things Without Touching Them

Read the text. **Think** about how the motion of a bowling ball changes after it starts rolling.

Read Together

Moving Things Without Touching Them

A ball is one object that can move. What happens when you roll a bowling ball? How does it stop?

Does a moving object only stop moving if it hits something else?

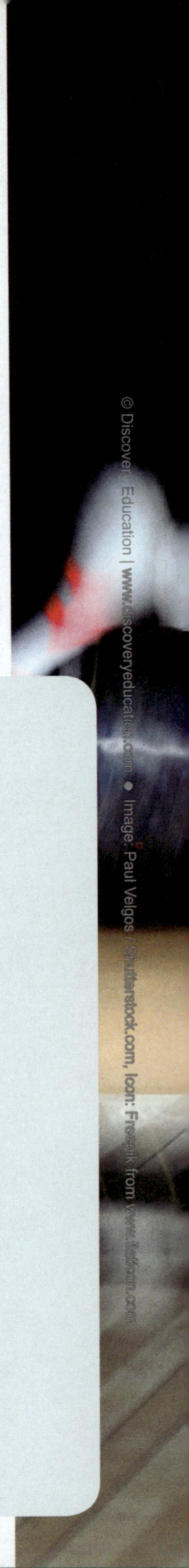

Activity 4

Observe Like a Scientist

Bowling

Watch the video. **Talk** about the question.

Bowling

Talk Together

How does a bowler change the motion of the ball?

Activity 5

Evaluate Like a Scientist

Quick Code: ca796s

What Do You Already Know About Controlling Motion?

What Do You Know?

A bowling ball rolls down a lane. How did the ball start rolling? What happens when it hits the pins? How will it stop moving? **Explain** using words or pictures.

More or Less Friction?

Which of the pictures show more **friction**? Which show less friction? **Draw** a line from each picture to the correct term.

Ice Skates

Boat on Sand

Dry Road

More Friction | **Less Friction**

Wet Road

Inline Skates

Boat in Water

Changing the Direction

Look at the picture. **Think** about how the boy controls where the ball goes.

Basketball Fun

A boy pushes a basketball down toward the ground. What will happen to the **direction** the ball is moving when it meets the ground? How is the boy controlling the ball's motion? **Explain** using words or pictures.

Controlling Motion

Look at the picture. **Think** about how the wall helps control the ball.

Playing Ball

When children play with a ball on a field, it can get loose. It might go into the street. What was built to control the ball so it doesn't go out of the yard? How does this control the motion of the ball? **Explain** using words or pictures.

Read Together

Speed Bump

A car can move fast over a smooth surface. A speed bump has a rough surface. It slows the car down.

Friction happens when objects rub against each other. Friction is a kind of force. A force is a push or a pull.

Speed Bump

What Is Needed to Make Something Go Faster or Slower?

Activity 6

Analyze Like a Scientist

Quick Code: ca800s

Speed Bump

Think about what you read. Then, **write** or **draw** your answer.

What makes cars slow down when they move across speed bumps?

CCC Cause and Effect

Activity 7

Observe Like a Scientist

Quick Code: ca802s

Fast and Slow

Watch the video. **Use** the graphic organizer to record what made the cars go faster or slower.

Fast and Slow

SEP Constructing Explanations and Designing Solutions

CCC Cause and Effect

Car	What Happened?	How Car Moved

Activity 8

Investigate Like a Scientist

Quick Code:
ca803s

Hands-On Investigation: Rough and Smooth

In this activity, you will observe how objects move over different surfaces.

Make a Prediction

How will an object move over a smooth surface? How will an object move over a rough surface? **Write** or **draw** your answers.

SEP Planning and Carrying Out Investigations

CCC Cause and Effect

CCC Patterns

What materials do you need?

(per group)

- Wooden board
- Wood block
- Marbles, 1 in.
- Bouncy balls
- Toy trucks
- Hand lens
- Felt
- Sandpaper
- Bubble wrap
- Masking tape
- Plastic wrap

HANDS-ON INVESTIGATION

What Will You Do?

Tape one material on the ramp. **Raise** the ramp on blocks. **Draw** your setup below.

Roll different objects down the ramp.

Repeat with different materials. **Look** carefully to see what happens.

What was the roughest surface?

What was the smoothest surface?

Which surfaces make objects speed up?

Which surfaces make objects slow down?

Record what you saw in the table.

Speed Up	Slow Down

Fill in the chart with words, pictures, or both.

	What does the surface look like up close?	Faster, slower, or same speed?	
		Material at Bottom of Ramp	Material on Ramp
Ramp with No Material			
Material 1			
Material 2			

Think About the Activity

Complete the sentence frames to describe the pattern you saw.

Surfaces that looked and felt rough made things

go ________________.

A ________________ surface would be easiest to move across.

Activity 9

Observe Like a Scientist

Quick Code: ca804s

Moving Slower

Watch the video. **Think** about how different surfaces change motion.

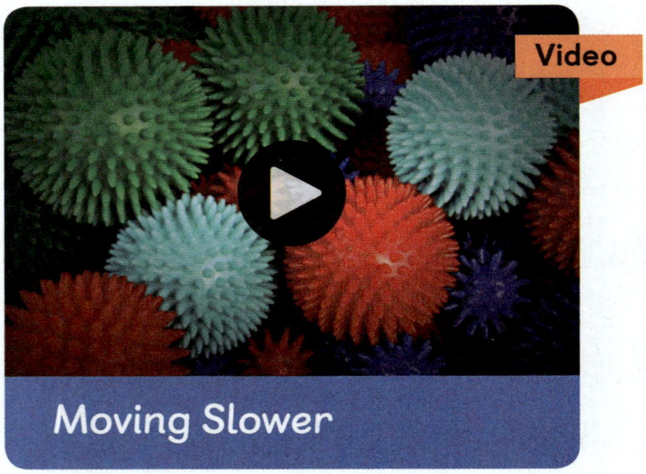

Moving Slower

Talk Together

Talk together about how friction makes things slow down.

SEP Obtaining, Evaluating, and Communicating Information

CCC Cause and Effect

Activity 10

Observe Like a Scientist

Quick Code: ca805s

Friction

Go online to **explore** how friction changes motion using the online activity. **Look** for actions that speed up or slow down motion.

Friction Exploration

SEP Developing and Using Models

CCC Cause and Effect

Write down actions that speed up or slow down the motion of the wooden box.

Speed Up Motion

Slow Down Motion

Activity 11

Observe Like a Scientist

Quick Code: ca806s

What Is Friction?

Watch the video. Look for actions that speed up or slow down motion.

What Is Friction?

SEP Obtaining, Evaluating, and Communicating Information

CCC Cause and Effect

Write down actions that speed up or slow down motion in the table.

Speed Up Motion

Slow Down Motion

How Can We Change Where a Moving Thing Goes?

Activity 12

Observe Like a Scientist

Quick Code:
ca807s

Changes in Motion

Watch the video. **Look** for examples of objects changing direction.

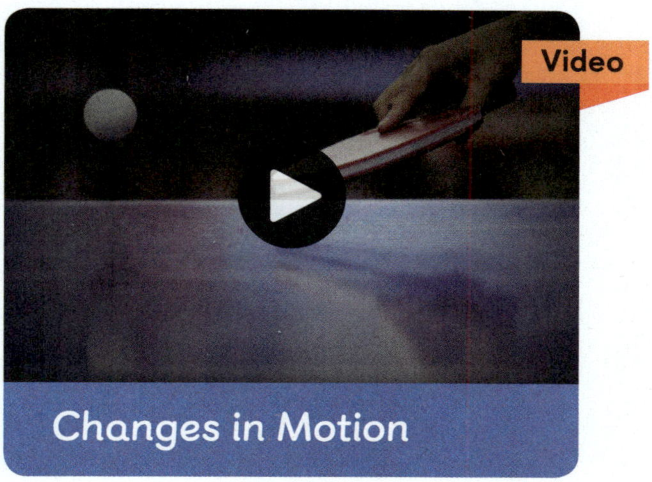

SEP Obtaining, Evaluating, and Communicating Information

CCC Cause and Effect

Talk Together

Talk together about how to change the motion and direction of a ball.

Activity 13

Analyze Like a Scientist

Quick Code: ca808s

Collisions

Read the text, and then **answer** the question.

Read Together

Collisions

Imagine a softball game. A pitcher throws the ball. The batter swings. Her bat hits the softball. The ball flies out into the field.

When one object hits another, they collide. In the softball game, the ball and bat collide with each other.

SEP **Obtaining, Evaluating, and Communicating Information**

Read Together

An object in motion can change its motion when it collides with another object.

Playing Cricket

For example, the direction of the ball changed when it collided with the bat. The force of the swinging bat pushed the ball out of the park.

Look at the picture. What is going to happen when the ball hits the bat?

Activity 14

Investigate Like a Scientist

Quick Code: ca810s

Hands-On Investigation: Colliding Objects

In this activity, you will look at what happens when a marble collides with different things.

Make a Prediction

What do you think will happen to a marble when it hits another marble? **Write** or **draw** your prediction.

What materials do you need?

(per group)

- Tube, cardboard, toilet paper roll
- Marbles, 1 in.
- Metric ruler

HANDS-ON INVESTIGATION

What Will You Do?

Roll a marble down the track so it hits another marble. **Roll** a marble against a barrier. **Roll** marbles so they hit each other. **Tell** what happened.

SEP Planning and Carrying Out Investigations

CCC Cause and Effect

Fill in the chart using words, pictures, or both.

Observations

What does the marble hit?	What happens to the marble?	What happens to the object the marble hits?

Think About the Activity

What happened to the speed of the marble when it hit something?

What happened to the object that got hit by the marble?

What happened to the direction of the marble when it hit something?

Activity 15

Observe Like a Scientist

Quick Code: ca811s

Moving Hockey Puck

Watch the video. **Think** about how hockey players change the direction of the puck during a game.

Talk Together

When have you changed the direction of an object in gym class?

Read Together

Controlling a Ball

We can use force to control the direction an object moves.

Girl Rolling a Ball

The rubber tires on a bicycle keep it going forward and not sideways.

A soccer player can kick a ball toward the goal.

A basketball can bounce off the backboard into the net.

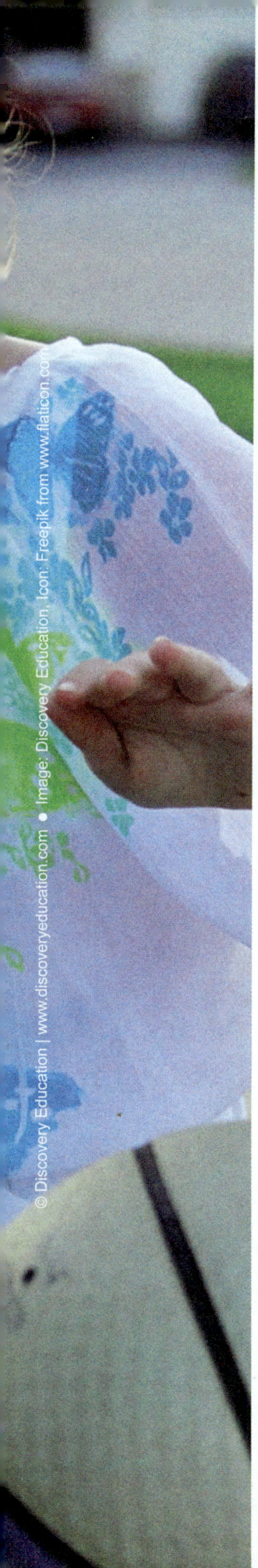

What Problems Can Be Fixed Using Force and Motion?

Activity 16

Analyze Like a Scientist

Quick Code: ca812s

Controlling a Ball

Think about what you read. Then **write** or **draw** your answer.

How can the girl make the ball stay on the sidewalk when it rolls?

Activity 17

Observe Like a Scientist

Quick Code: ca814s

Changing Things with Force

Watch the video. **Look** for how force is used to change the direction of a moving object.

Video

Changing Things with Force

Fill in the bubble map to answer the question in the middle bubble.

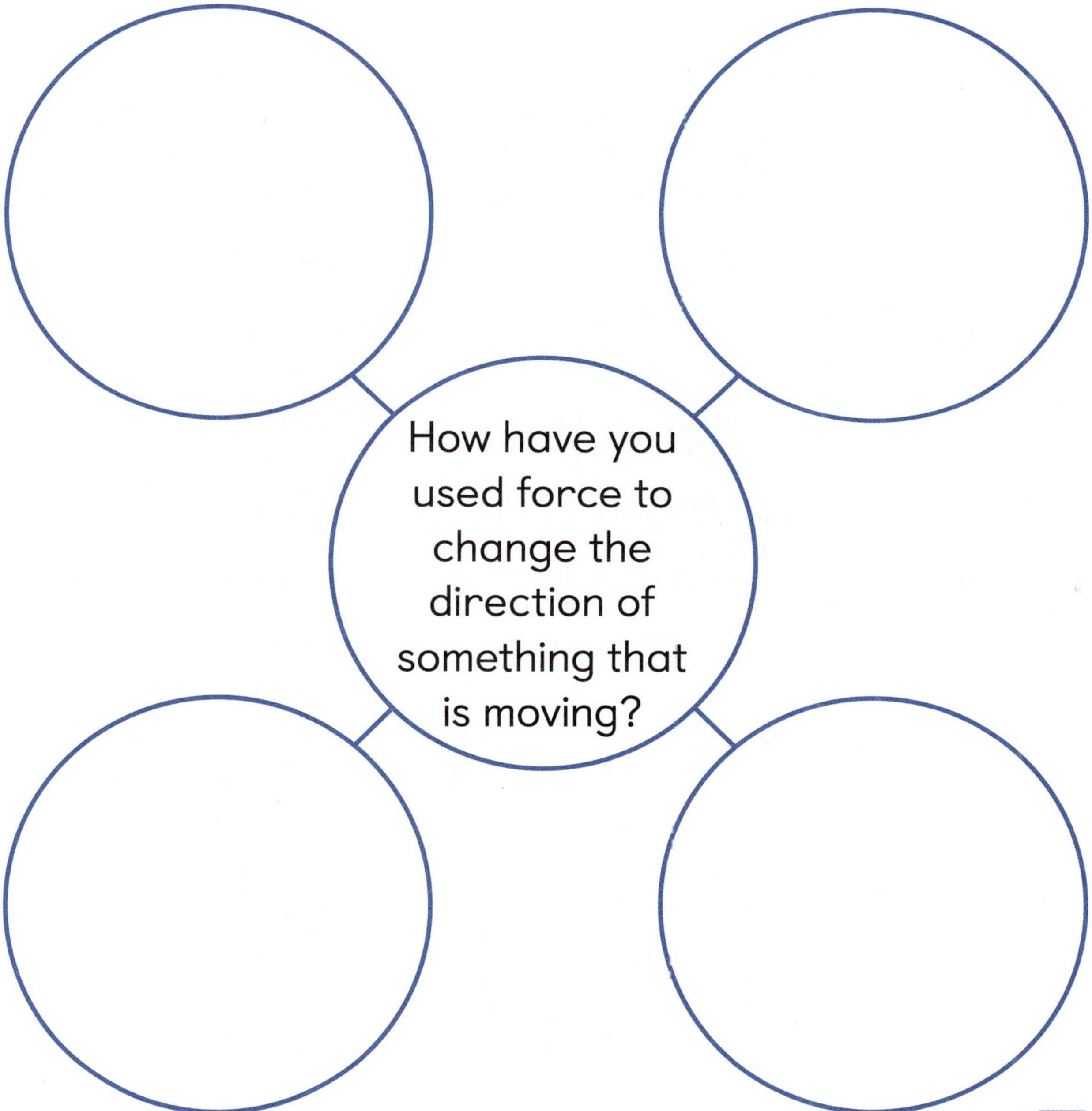

Activity 18

Design Solutions Like a Scientist

Quick Code: ca815s

Hands-On Engineering: Controlling the Direction Game

In this activity, you will design and create a rolling marble game. The game should keep the marble moving inside a box as long as possible with a single push.

Ask Questions About the Problem

How will you keep the marble moving? What questions do you have before you begin your design?

SEP Constructing Explanations and Designing Solutions

CCC Cause and Effect

What materials do you need?

(per group)

- Cardboard box, shallow (paper box lid or shirt box)
- Several small wooden blocks
- Markers
- Cardstock
- Masking tape
- Image of a pinball machine
- Marbles, 1 in.
- Pencils

HANDS-ON ENGINEERING

What Will You Do?

Make a game that keeps a marble moving inside a cardboard box for as long as possible. **Test** your design as you go. **Make** changes to improve the design.

Write or **draw** the design of your game.

Think About the Activity

What caused the marble to travel inside the box?

How did the materials keep the marble moving inside the box?

What could you do to your design to keep the marble in the box longer?

Draw how you would change your game design so the marble moves longer.

Read Together

Stopping Motion

Sometimes, we want to make sure an object does not move. Can you think of objects we do not want to move if someone or something pushes or pulls on them?

Bicycling

When forces are **balanced**, an object does not move. We do not want a house to move. We build houses to balance the forces that act on them. A house must not fall down. It must not move in the wind.

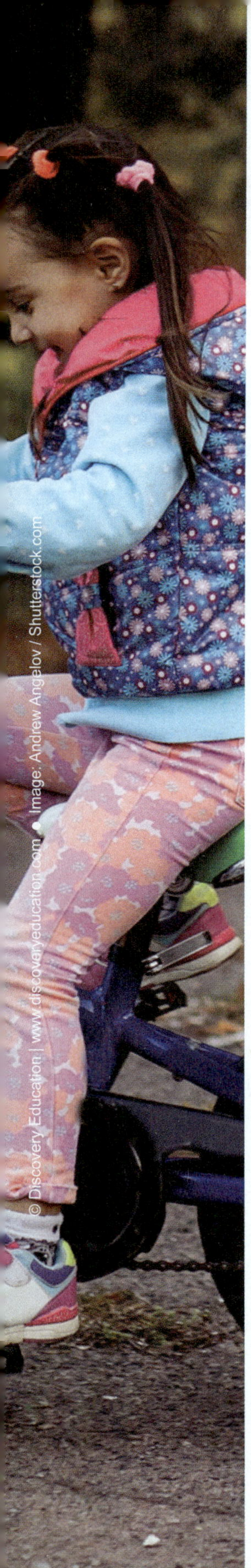

Activity 19

Analyze Like a Scientist

Quick Code: ca816s

Stopping Motion

Think about what you read. Then, **answer** the question.

How do balanced forces help the boy stay on the bicycle?

SEP Asking Questions and Defining Problems

CCC Cause and Effect

Activity 20

Observe Like a Scientist

Quick Code: ca818s

Balanced Forces

Watch the video. **Think** about how balanced forces prevent objects from moving.

Balance

Play the animation. Then, **answer** the questions.

Balanced Forces

What happens when two forces are balanced?

What happens when one force is stronger than another?

SEP **Constructing Explanations and Designing Solutions**

Activity 21

Design Solutions Like a Scientist

Quick Code: ca820s

Hands-On Engineering: Stay Standing

In this activity, you will design and create a tower. You will then try to protect your tower from a wrecking ball.

Ask Questions About the Problem

What do you need to know before you design your tower?

SEP Planning and Carrying Out Investigations

CCC Structure and Function

What materials do you need?
(per group)

- Tennis ball
- String
- Sugar cubes

HANDS-ON ENGINEERING

What Will You Do?

Write or **draw** what your tower will look like.

Build your tower using the stackable materials and cardboard sheets. **Test** your design as you build. **Make** changes to improve the design. **Use** the wrecking ball to try to knock down your tower.

Write or **draw** a new design for your tower.

Think About the Activity

How did gravity affect your tower?

What is the best material to use to keep the ball from knocking down your tower?

Activity 22

Record Evidence Like a Scientist

Quick Code: ca821s

A Moving Bowling Ball

Now that you have learned about controlling motion, look again at the Moving Bowling Ball image. You first saw this in Wonder.

Let's Investigate a Moving Bowling Ball

Talk Together

How can you describe A Moving Bowling Ball now? How is your explanation different from before?

SEP Constructing Explanations and Designing Solutions

Look at the Can You Explain? question. You first read this question at the beginning of the lesson.

Can You Explain?

How can we change the speed and direction of a bowling ball?

Now you will use your new ideas about a moving bowling ball to answer a question.

1. **Choose** a question. You can use the Can You Explain? question, or one of your own. You can also use one of the questions that you wrote at the beginning of the lesson.

My Question

2. Now, **use** the sentence frames to answer the question.

When the bowling ball moved forward, it was because a force

The bowling ball rolled in one direction until

The ball stopped because

STEM in Action

Activity 23

Analyze Like a Scientist

Quick Code: ca824s

Astronauts at Work

Read the passage and **watch** the video.

Look at the picture. **Find** differences between Earth and the moon.

Answer the question that follows.

SEP Obtaining, Evaluating, and Communicating Information

Read Together

Astronauts at Work

Have you ever dropped a book, and it fell to the ground? Have you ever seen leaves blow around in the fall? Different forces act on us every day. Astronauts have to understand how many forces work all the time.

Video

Wind

Being an astronaut can be difficult. There are many different forces to understand. Wind is one force astronauts must understand. They must also understand a force called *gravity*.

When you drop something, it falls toward Earth. Gravity is the force that causes objects to fall toward Earth's center.

Wind pushes rockets around. Gravity pulls the International Space Station toward Earth.

A rocket going into space needs to fight against the force of gravity to fly away from Earth.

Kicking Around

Kicking Around

What would it be like to play soccer on the moon? **Draw** or **write** your answer.

Activity 24

Evaluate Like a Scientist

Quick Code: ca827s

Review: Controlling Motion

Think about what you have read and seen. What did you learn?

Draw what you have learned. Then **tell** someone else about what you learned.

Talk Together

Think about what you saw in Get Started. Use your new ideas to discuss Controlling Motion.

SEP Obtaining, Evaluating, and Communicating Information

Design Solutions Like a Scientist

Quick Code: ca834s

Hands-On Engineering: Investigating Motion

In this activity, you will design a solution that will help you move a heavy box from a low place to a high place.

Moving Boxes

SEP Asking Questions and Defining Problems

SEP Planning and Carrying Out Investigations

SEP Constructing Explanations and Designing Solutions

CCC Cause and Effect

What materials do you need?
(per group)

- Heavy box (e.g., a shoebox containing one or two bricks)
- Wooden board
- Rope
- Cloth, nylon
- Plastic wrap
- Felt
- Sandpaper

HANDS-ON ENGINEERING

Ask Questions About the Problem

Think about what you learned about the motion of an object.

What materials will help the box move? How will you move your box?

What Will You Do?

Make your model. How will you know if it will work?

Test your design. **Draw** pictures to show how you will move the box.

Think About the Activity

Write or **draw** your answers to the questions in the chart.

How well did your box move?

How could you improve your design?

What Worked?	What Didn't Work?
What Could Work Better?	

Grade K Resources

- Bubble Map
- Safety in the Science Classroom
- Vocabulary Flash Cards
- Glossary
- Index

Name ______________________________

Bubble Map

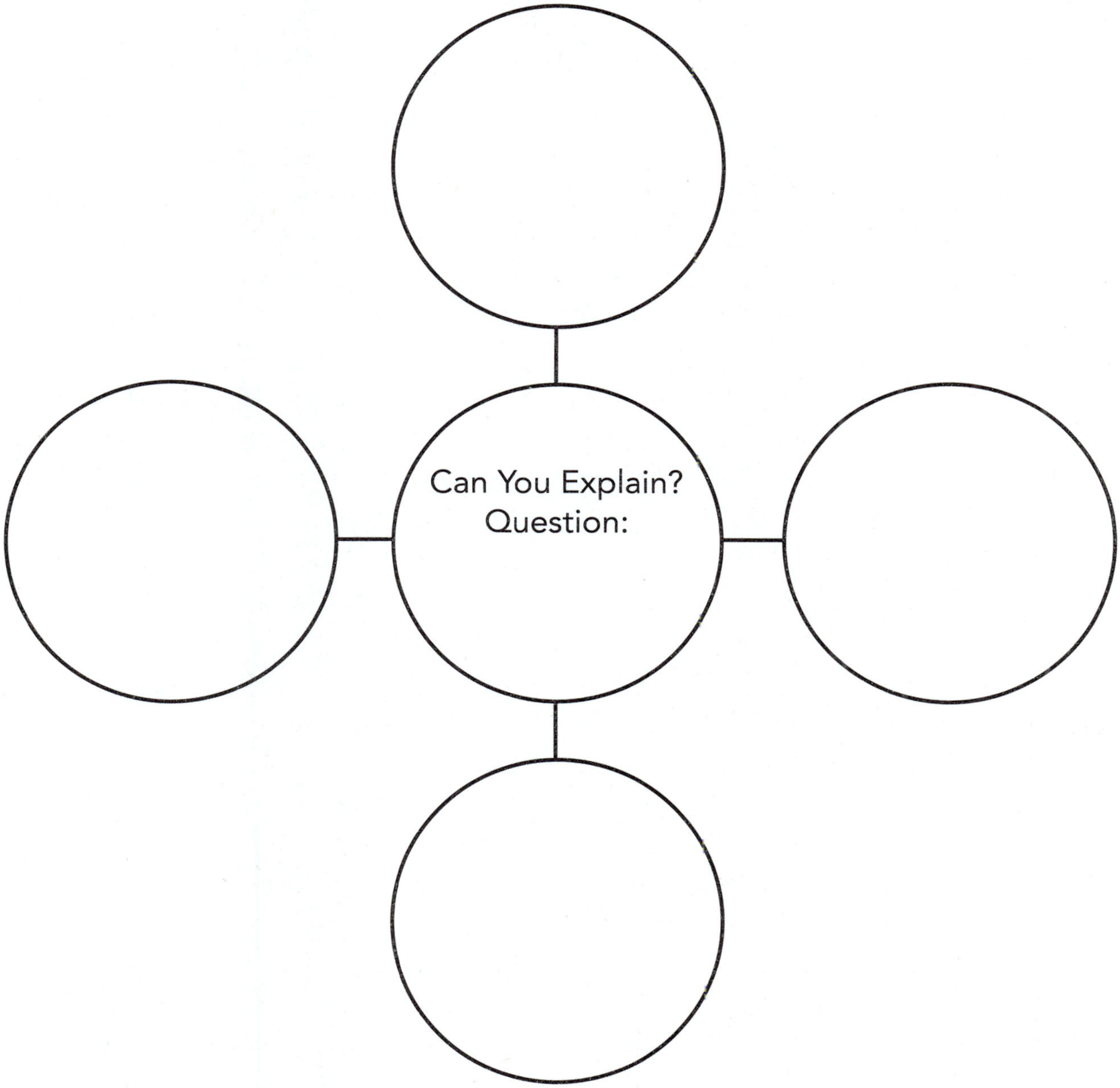

Following common safety practices is the first rule of any laboratory or field scientific investigation.

Dress for Safety

One of the most important steps in a safe investigation is dressing appropriately.

- Splash goggles need to be kept on during the entire investigation.
- Use gloves to protect your hands when handling chemicals or organisms.
- Tie back long hair to prevent it from coming in contact with chemicals or a heat source.

Safety Goggles

- Wear proper clothing and clothing protection. Roll up long sleeves, and if they are available, wear a lab coat or apron over your clothes. Always wear closed-toe shoes. During field investigations, wear long pants and long sleeves.

Be Prepared for Accidents

Even if you are practicing safe behavior during an investigation, accidents can happen. Learn the emergency equipment location in your classroom and how to use it.

- The eye and face wash station can help if a harmful substance or foreign object gets into your eyes or onto your face.

- Fire blankets and fire extinguishers can be used to smother and put out fires in the laboratory. Talk to your teacher about fire safety in the lab. He or she may not want you to directly handle the fire blanket and fire extinguisher. However, you should still know where these items are in case the teacher asks you to retrieve them.

Most importantly, when an accident occurs, immediately alert your teacher and classmates. Do not try to keep the accident a secret or respond to it by yourself. Your teacher and classmates can help you.

Fire Extinguisher

Practice Safe Behavior

There are many ways to stay safe during a scientific investigation. You should always use safe and appropriate behavior before, during, and after your investigation.

- Read all of the steps of the procedure before beginning your investigation. Make sure you understand all the steps. Ask your teacher for help if you do not understand any part of the procedure.
- Gather all your materials and keep your workstation neat and organized. Label any chemicals you are using.
- During the investigation, be sure to follow the steps of the procedure exactly. Use only directions and materials that have been approved by your teacher.

- Eating and drinking are not allowed during an investigation. If asked to observe the odor of a substance, do so using the correct procedure known as wafting, in which you cup your hand over the container holding the substance and gently wave enough air toward your face to make sense of the smell.
- When performing investigations, stay focused on the steps of the procedure and your behavior during the investigation. During investigations, there are many materials and equipment that can cause injuries.
- Treat animals and plants with respect during an investigation.

- After the investigation is over, appropriately dispose of any chemicals or other materials that you have used. Ask your teacher if you are unsure of how to dispose of anything.
- Make sure that you have returned any extra materials and pieces of equipment to the correct storage space.
- Leave your workstation clean and neat. Wash your hands thoroughly.

balanced	direction
 mage: Olesia Bilkei / Shutterstock.com	 Image: sergign / Shutterstock.com
when something has equal weight or force on both sices	the line in which a person or object is moving

energy	engineer
 Image: Discovery Education	 Image: Pixabay
the ability to do work or make something change	a person who designs something that may be helpful to solve a problem

force

Image: Nasa

a push or a pull on an object

friction

Image: Corepics VOF / Shutterstock.com

a force that happens when two objects rub together

function

Image: Pixabay

purpose

gravity

Image: Mauricio Graiki / Shutterstock.com

the force that pulls an object toward the center of Earth

living

Image: Comstock / Stockbyte / Getty Images

things that are alive, such as people and animals

motion

Image: Comstock / Stockbyte / Getty Images

going from one place to another

nonliving

Image: Pixabay

things that are not alive, such as a ball or block

English — A — Español

English	Español
air an invisible gas that is all around us; living things, such as plants and animals, need it to breathe and grow	**aire** gas invisible que nos rodea; todos los seres vivos, como las plantas y los animales, lo necesitan para respirar y crecer
animal a living thing that moves around to look for food, water, or shelter, but can't make its own food	**animal** ser vivo que se mueve para buscar alimento, agua o refugio, pero no puede producir su propio alimento

— B —

English	Español
balanced when something has equal weight or force on both sides	**equilibrado** cuando algo tiene un peso o fuerza igual en ambos lados

basic need
a most important thing needed to survive, such as food, water, and shelter

necesidad básica
lo más importante que se necesita para sobrevivir, como alimento, agua y refugio

---

bird
an animal with feathers, wings, and a beak whose babies are born from eggs

ave
animal que tiene plumas, alas y un pico cuyos bebés nacen de huevos

---

blizzard
a very big snowstorm

ventisca
tormenta de nieve muy grande

C

chemical
created when certain substances are mixed together

sustancia química
se forma cuando se mezclan otras sustancias

cloud
a group of water droplets or ice crystals that form together in the sky

nube
grupo de gotitas de agua o cristales de hielo que se forman en el cielo

---

compare
to look at two or more things to see what is the same and different

comparar
observar dos o más cosas para ver en qué se parecen y se diferencian

D

data
information that is used to learn about something

dato
información que se usa para aprender acerca de algo

---

describe
to tell or write about something, like creating a picture with words

describir
contar o escribir acerca de algo, como crear un dibujo con palabras

design
a plan of how to create or build something

diseño
plan de cómo crear o construir algo

---

direction
the line in which a person or object is moving

dirección
línea en la que se mueve una persona u objeto

E

energy
the ability to do work or make something change

energía
habilidad de trabajar o producir un cambio

---

engineer
a person who designs something that may be helpful to solve a problem

ingeniero
persona que diseña algo que puede ser útil para resolver un problema

environment
all the living and nonliving things that surround an organism

medio ambiente
todos los seres vivos y objetos sin vida que rodean a un organismo

F

food
what living things need to eat so they can grow and survive

alimento
lo que los seres vivos necesitan comer para crecer y sobrevivir

force
a push or a pull on an object

fuerza
acción de atraer o empujar que se aplica a un objeto

forecast
a prediction of what the weather will be like

pronóstico
predicción sobre cómo será el tiempo

friction
a force that happens when two objects rub together

fricción
fuerza que ocurre cuando se frota un objeto con otro

---

function
purpose

función
propósito

G

gas
a liquid made from crude oil and used as a fuel for objects, such as cars

gasolina
líquido hecho a partir del petróleo que se usa como combustible en algunos objetos, como los carros

---

grassland
a large area of land covered by grass

pradera
gran área de tierra cubierta de pasto

gravity
the force that pulls an object toward the center of Earth

gravedad
fuerza que tira de un objeto hacia el centro de la Tierra

---

grow
to get larger

crecer
hacerse más grande

H

habitat
the place where a plant or animal lives

hábitat
lugar donde vive una planta o un animal

---

hurricane
a big storm with strong winds and a lot of rain

huracán
gran tormenta con vientos fuertes y mucha lluvia

I

information
facts or knowledge that come from a source, such as a book or a person

información
datos o conocimientos que provienen de una fuente, como un libro o una persona

L

landscape
the view of a land's surface

paisaje
vista de la superficie de un terreno

---

light
a form of energy that makes it possible for our eyes to see

luz
forma de energía que hace posible ver con los ojos

---

living
things that are alive, such as people and animals

vivo
seres que viven, como las personas y los animales

M

material
things that can be used to build or create something

material
cosas que se pueden usar para construir o crear algo

---

measure
to find the amount, the weight, or the size of something

medir
hallar la cantidad, el peso o el tamaño de algo

---

measurement
information that tells us about the amount, the weight, or the size of something

medida
información que nos dice sobre la cantidad, el peso o el tamaño de algo

---

metal
a shiny material that can be used to make tools and machines

metal
material brillante que se puede usar para hacer herramientas y máquinas

meteorologist
a person who studies weather

meteorólogo
persona que estudia el tiempo

---

mineral
something that is formed in the earth and is not part of a plant or animal

mineral
algo que se forma en la tierra y no es parte de una planta o animal

---

model
a human-made version created to show the parts of something else, either big or small

modelo
versión creada por el hombre para mostrar las partes de algo más, ya sea grande o pequeño

---

motion
going from one place to another

movimiento
ir de un lugar a otro

N

natural resource
a material that comes from Earth and is found in nature

recurso natural
material que proviene de la Tierra y se encuentra en la naturaleza

nonliving
things that are not alive, such as a ball or block

sin vida
cosas que no están vivas, como una pelota o un bloque

O

observe
to watch closely

observar
mirar atentamente

organism
a living thing

organismo
ser vivo

P

plant
a living thing made up of cells that needs water and sunlight to survive

planta
ser vivo formado por células que necesita agua y luz solar para sobrevivir

---

pollution
when garbage, smells, or other materials are scattered over places that used to be clean

contaminación
cuando la basura, el olor u otros materiales se dispersan en lugares que solían ser limpios

---

predict
to make a guess based on what you already know

predecir
adivinar en base a lo que ya se sabe

R

record
to write down or make note of something

anotar
escribir o apuntar algo

recycle
to create new materials from something used

reciclar
crear nuevos materiales a partir de algo usado

---

reduce
to use less of something

reducir
usar menos de algo

---

renewable
to make new again

renovable
hacer que algo sea nuevo otra vez

---

reuse
to use something again

reutilizar
usar algo de nuevo

---

root
the part of a plant that usually grows under the ground

raíz
parte de una planta que por lo general crece bajo tierra

S

season a time of year: summer, autumn, winter, or spring	**estación** época del año: el verano, el otoño, el invierno o la primavera
shelter a place that protects you from harm or bad weather	**refugio** lugar para protegerse de peligros o el mal tiempo
soil dirt that covers Earth, in which plants can grow and insects can live	**suelo** tierra que cubre nuestro planeta en la que pueden crecer plantas y vivir insectos
study a close look at something	**estudio** vistazo detallado de algo

sun

any star around which planets revolve

sol

toda estrella alrededor de la cual giran los planetas

---

survive

to continue to live

sobrevivir

continuar viviendo

T

temperature

the measure of how hot or cold something is

temperatura

medida de lo caliente o frío que algo está

---

thermometer

a tool used to measure temperature

termómetro

instrumento que se usa para medir la temperatura

tornado
a storm that creates a funnel cloud with extremely fast winds

tornado
tormenta que crea una nube con forma de embudo que tiene vientos extremadamente rápidos

V

vegetable
a plant or part of a plant that is used for food

vegetal
planta o parte de una planta que se usa como alimento

W

warm
having heat

cálido
que tiene calor

waste
garbage or other materials you don't need

deshecho
basura u otros materiales que no necesitamos

water a clear liquid that has no taste or smell	**agua** líquido transparente que no tiene sabor ni olor
weather the state of the outdoors at a given time and place	**tiempo atmosférico** el estado del exterior en un momento y lugar dados
wildlife plants and animals that live in nature	**vida silvestre** plantas y animales que viven en la naturaleza

Index

D

E

N

O

P

R

S

T

W